BIG CATS

TIGERS

by Elizabeth Andrews

Cody Koala
An Imprint of Pop!
popbooksonline.com

Hello! My name is Cody Koala

This book is filled with videos, puzzles, games, and more! Scan the QR codes* while you read, or visit the website below to make this book pop.

popbooksonline.com/tiger

*Scanning QR codes requires a web-enabled smart device with a QR code reader app and a camera.

abdobooks.com
Published by Pop!, a division of ABDO, PO Box 398166, Minneapolis, Minnesota 55439.

Printed in the United States of America, North Mankato, Minnesota.

102024
012025

Cover Photo: Shutterstock Images
Interior Photos: Getty Images, Shutterstock Images
Editor: Grace Hansen
Series Designer: Neil Klinepier, Candice Keimig

Library of Congress Control Number: 2024938602

Publisher's Cataloging-in-Publication Data
Names: Andrews, Elizabeth, author.
Title: Tigers / by Elizabeth Andrews
Description: Minneapolis, Minnesota : Pop!, 2025 | Series: Big cats | Includes online resources and index
Identifiers: ISBN 9781098246938 (lib. bdg.) | ISBN 9781098247492 (ebook)
Subjects: LCSH: Big cats--Juvenile literature. | Wildcat--Juvenile literature. | Tigers--Juvenile literature. | Tiger--Behavior--Juvenile literature.
Classification: DDC 599.756--dc23

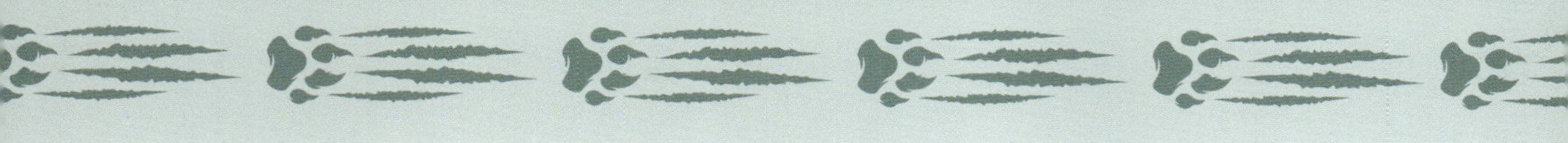

Table of Contents

Chapter 1

What Is a Tiger?

The tiger is the largest big cat. Tigers have thick orange fur with dark stripes. No two tigers have the same stripe pattern. Their bellies are white.

Tigers can weigh more that 650 pounds (295kg).

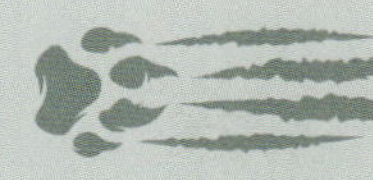

Watch a video here!

Chapter 2

Where Do Tigers Live?

Tigers live throughout Asia. They live in lots of different **habitats**, including swamps, rainforests, and grasslands. Tigers like to live near water. They can swim!

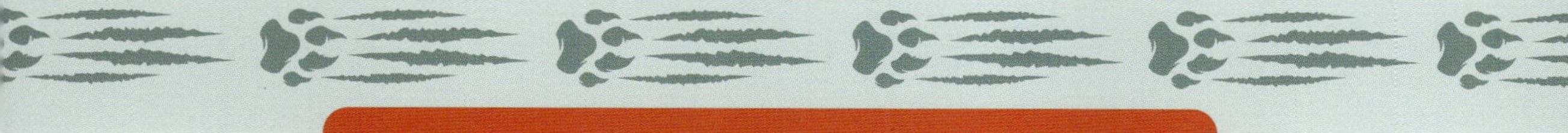

Where Tigers Live

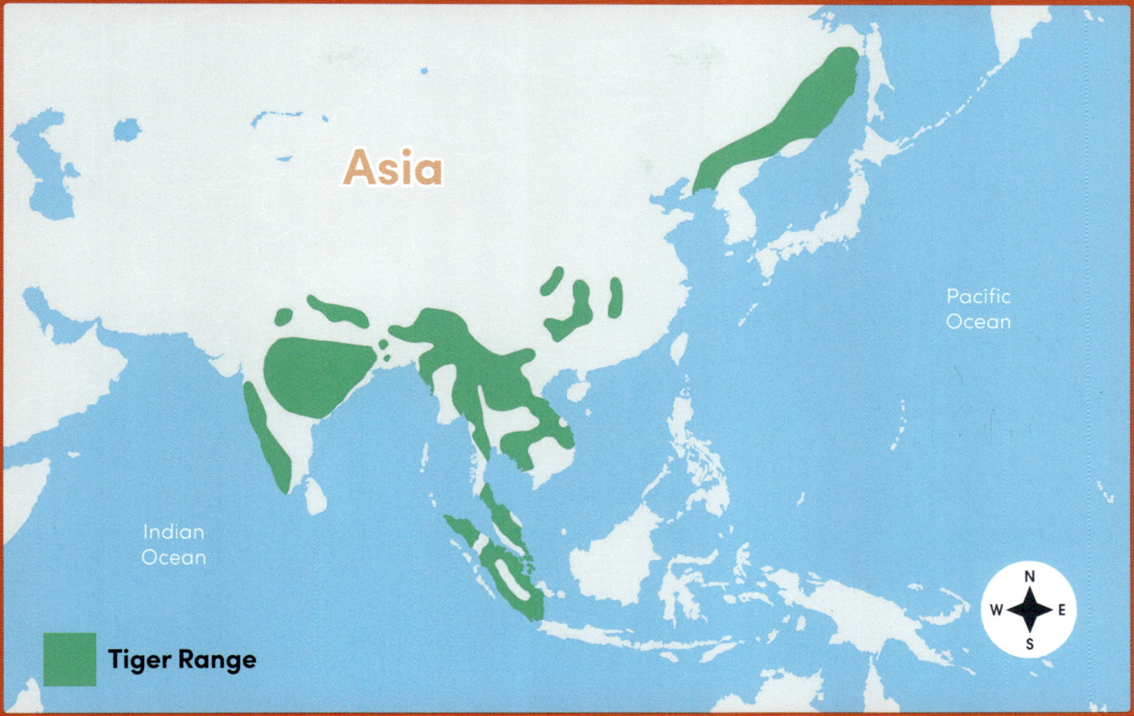

Asia

Pacific Ocean

Indian Ocean

N W E S

Tiger Range

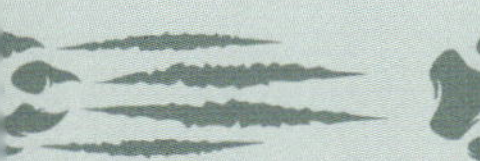

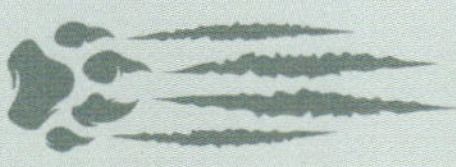

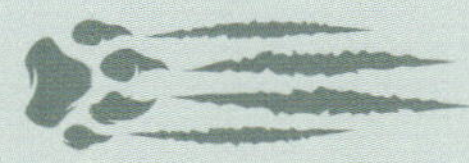

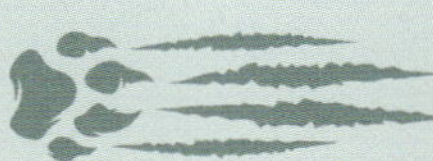

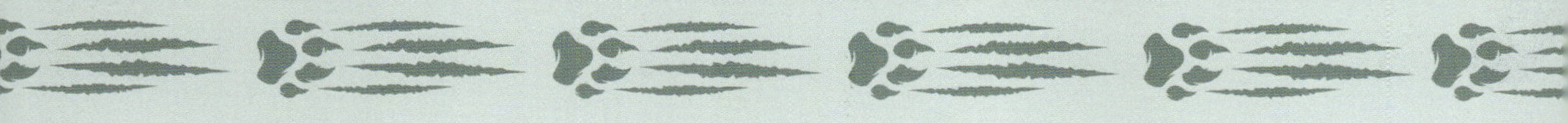

The largest tigers are called Amur tigers. They live in mountain forests where it gets very cold. Sumatran tigers are the smallest. They live in Indonesia.

Sumatran tigers only grow to be about 300 pounds (136kg).

Chapter 3

Tiger Habits

Tigers are **solitary** animals. They live very far away from one another. Tigers know where their **territory** ends and another begins.

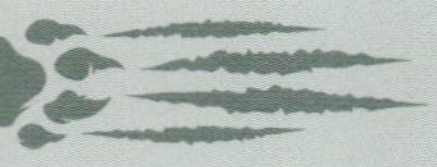
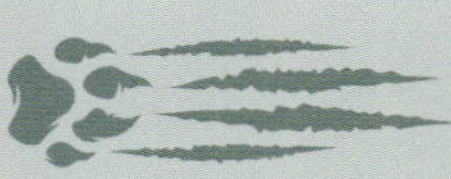
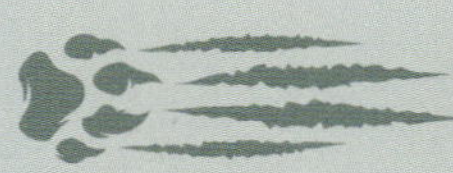
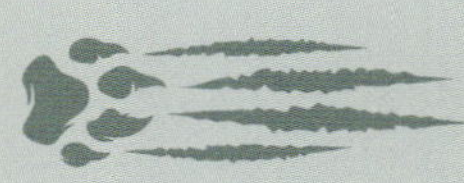

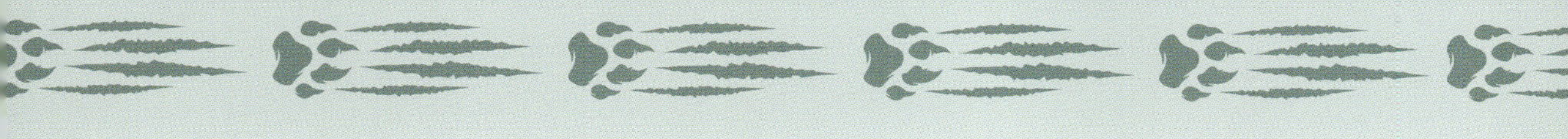

Explore links here!

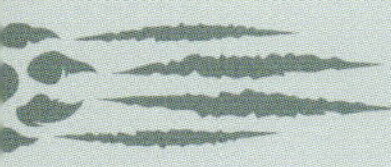

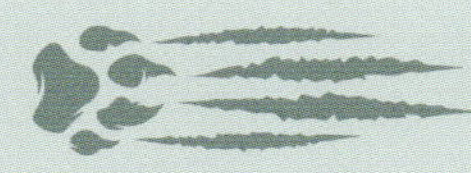
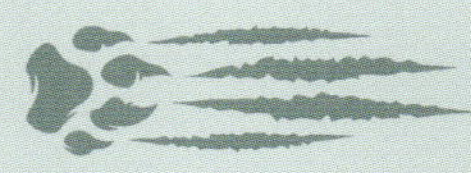
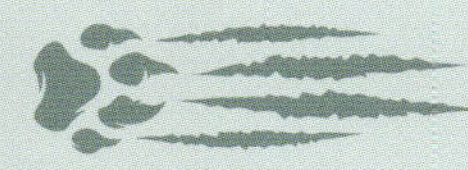

Tigers are very protective of their territory. They mark their territory by scratching trees and leaving their scent.

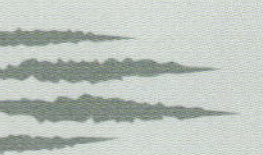
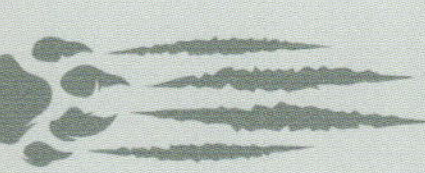
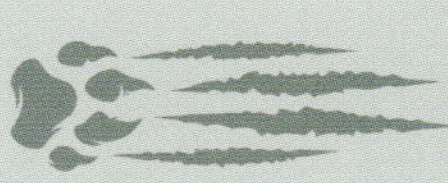

They communicate with each other using roars and grunts.

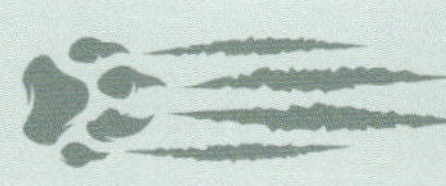

Tiger are **carnivores**. They eat animals, such as wild boar, deer, monkeys, and water buffalo. Tigers only hunt once a week. They can eat 80 pounds (36.3kg) of food in one day.

Tigers can hunt in water. They have been known to eat crocodiles!

Tigers often hunt at night. Their striped coats help them **camouflage**. Tigers quietly hide until their **prey** passes by. Then they quickly pounce and kill it.

Chapter 4

Tiger Cubs

Mother tigers have two to four cubs at a time. Cubs are born with their orange fur and dark stripes. They live in a den with their mother until they are six weeks old.

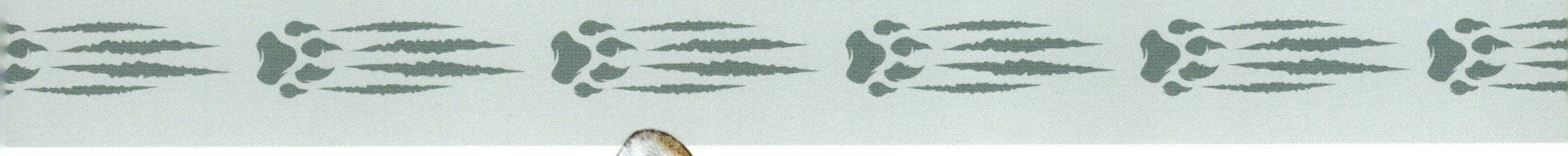

Complete an activity here!

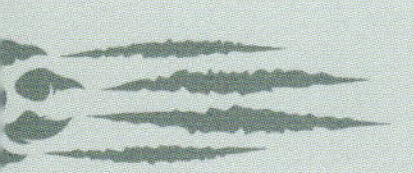

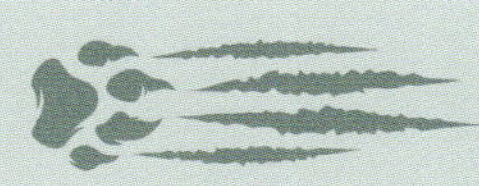
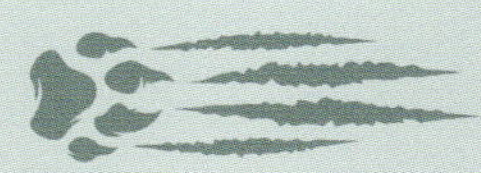
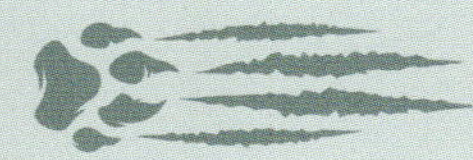

Cubs stay with their mother until they are two years old. They learn to hunt by watching her. Female tigers leave and find **territory** near their mother. Male tigers go farther from home.

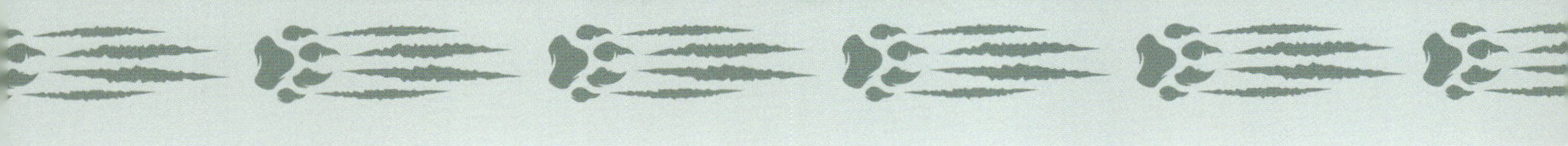

Tigers live around 10 to 15 years in the wild.

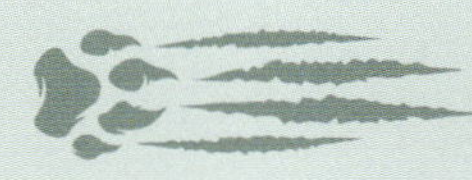
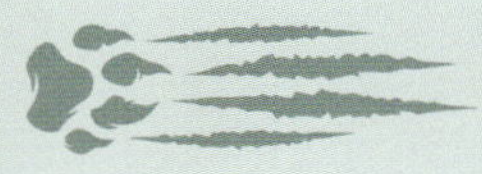

Making Connections

Text-to-Self

What big cat are you most interested in? Please explain your answer.

Text-to-Text

Have you read about any other kinds of big cats? If so, how were those big cats similar to or different from tigers?

Text-to-World

Tigers' stripes help them camouflage. Are there any other animals that have stripes for the same reason?

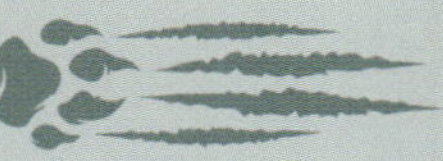

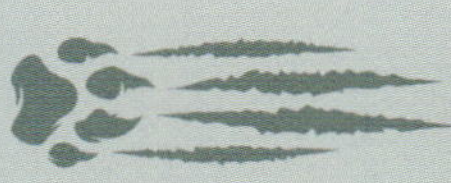
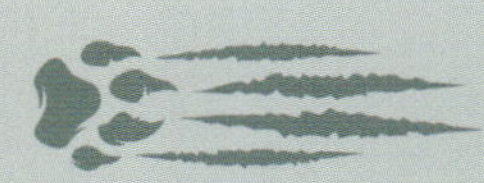
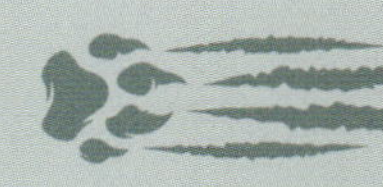

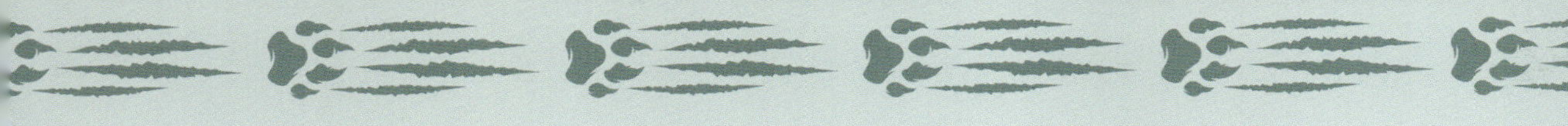

Glossary

carnivore – an animal that feeds on other animals.

camouflage – to hide by coloring or covering to look like the surroundings.

habitat – the home of an animal.

prey – an animal hunted by other animals for food.

solitary – living alone.

territory – a particular area of land that belongs to an animal.

Index

Online Resources

popbooksonline.com

Thanks for reading this Cody Koala book!

This book is filled with videos, puzzles, games, and more! Scan the QR codes* while you read, or visit the website below to make this book pop.

popbooksonline.com/tiger

*Scanning QR codes requires a web-enabled smart device with a QR code reader app and a camera.